D0948624

BUILDING MACHINES

EDITED WITH AN INTRODUCTION BY ROBERT MCCARTER

PRINCETON ARCHITECTURAL PRESS

PAMPHLET ARCHITECTURE NO. 12
BUILDING; MACHINES

Editor: Robert McCarter
Assistant Editor/Layout Designer: Julia Bourke

Cover & Frontispiece: Neil Denari
Editorial Advisors: Steven Holl, Ken Kaplan
Essay photographs: Christopher Scholz
Production: Ann Urban

Pamphlet Architecture is distributed exclusively by Princeton Architectural Press, 2 Research Way, Forrestal Center, Princeton, New Jersey 08540. Telephone: 609-987-2424.

ISBN 0-910413-40-1

T A B L E   O F   C O N T E N T S

fig. i. *"Like their towers, the men are dressed in costumes whose essential characteristics are similar; only their gratuitous features are different."*

fig. ii. *"There are no holds barred on design in this true development class."*

PAMPHLET ARCHITECTURE

Pamphlet Architecture was initiated in 1977 as an independent vehicle to criticize, question and exchange views. Each issue assembled by an individual author/architect. For information, pamphlet proposals, or contributions New York, New York 10011.

PAMPHLETS PUBLISHED

*Out of Print

Pamphlet Architecture No. 12 has been made possible by contributions from:

Michael W. Cox
Joe and Serena Fenton
Roger Ferri
Gisue Hariri
Mojgan Hariri
Hellmuth, Obata & Kassabaum, Inc.
Larry Rouch & Company
Gail A. Lindsey
Clevon Pran
Peter C. Pran
The Pace Collection
Kathleen Schneider
Henry Smith-Miller
Susana Torre

fig. iii. Steel mill.

FOREWORD

During the past several years, six young architects, drawn from around the United States, have been meeting in their Manhattan studios to exchange ideas about architecture—attempting to hammer out a vision, rather than merely another version of it.

Here in New York City, in an atmosphere mixed with both the jet-black smoke of smoldering subway fires and the white exhaust clouds of cooling tower condensers above, these six architects—Neil Denari, Ted Krueger, Ken Kaplan, Chris Scholz, Peter Pfau, and Wes Jones—underwent a creative combustion, producing an assembly of works that was exhibited at P.S. 1, a public art gallery in Long Island City, during the winter of 1986.

As a public document of this event, Pamphlet Architecture has published this "manual," not unlike those driver's operating manuals often found stuffed into the glove compartments of diesel truck cabs, rarely retrieved except to insert under a back tire, in order to free it from an oil-soaked pothole. It is our hope that this Pamphlet will serve a similar, timely function for those architects stuck in the asphalt depressions of current architectural discourse.

fig. iv. Wind tunnel.

CAPE FROM THE REVOLVING DOOR:
hitecture and the Machine

bert McCarter

one of the more revealing ironies of our time that the progres-
e development of modern technological and scientific thinking
s culminated—in contemporary architecture—in the reactionary,
uctive, and superficial use of historical forms. In order to
ose this situation effectively, as the following projects are
mpting to do, it is necessary not only to create other, more
propriate forms, but also, and far more important, to understand
d resist the propensity of utilitarian and economic determinism
transform architectural design into an instrument of an
clusively technological character. This determinism, which is
ed to define our modern world, displays a total indifference to
chitectural form: the calculating production of technology, as
idegger has noted, is *an act without an image*.

While the "style" that superficially characterizes such economi-
lly determined buildings is thus understood to be irrelevant, the
current "style" debates that rage within the profession and
ademia have nevertheless obscured the more fundamental
positions created by the dominance of technological thinking.
e architectural profession, mesmerized by the endless cycle of
shion and increasingly unable to define any common disciplinary
inciples, appears today almost wholly incapable of formulating
y kind of critical position—much less any true opposition.

Perhaps the only real hope for the formation of an effective
sistance lies on the periphery of the profession: in the marginal
actices that are almost inevitably unprofitable, where, despite the
ressing difficulties of personal finances, thinking about architec-
re tends to be less determined by purely economic evaluations
an is the case in the profession at large. There is an intensity to
e found along this edge and a greater willingness to question the
tatus quo, to search for a more original understanding of architec-
ure and its relation to technology than can be found in the more
omfortable middle ground. It is here, perhaps, that one may find
e kind of thought and discussion that results in the formation of
ulture, as it is here that a resolution of the seemingly insoluable
pposition between the technological and the architectural is being
ttempted.

In Le Corbusier's *Vers une architecture*, certainly the best
known argument for a modern, machine-inspired architecture,
here already exists this opposition between that which is determined by economic and utilitarian evaluation and that which is
determined by an inherited architectural order. In this, the book is
almost schizophrenic—alternating chapters championing first the
engineer, the rational process of selection by economic criteria,
the house as "a machine for living in," and then the architect, the
creation of forms of a spiritual order, "the masterly, correct and
magnificent play of masses brought together in light." While Le
Corbusier in the end attempted to assert the dominance of the historical discipline of architecture, technological thinking was in fact
increasingly dominant in the definition of the modern world—and in
the definition of architecture as it was to exist in that world.

fig. v. Wheel factory.

fig. vi. Lightweight helicopter.

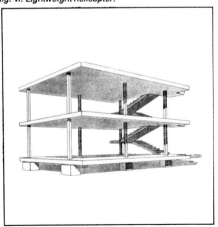

fig. vii. Le Corbusier: Maison Domino, 1914.

fig. viii. Treadmill at Brixton Prison.

fig. ix. Airplane.

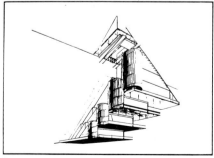

fig. x. Student project for restaurant, Vkhutemas.

Technological thinking, as Heidegger has found, is characte ized by the intention of controlling and dominating nature. Su thinking also controls and dominates man inasmuch as he is pa of nature and himself has a nature. Efficiency, economy, and u ity become ends in themselves rather than means to some oth end. Being, in this kind of thinking, consists of being *usable*.

In the same way that nature is a ceaseless process, technolo ical production is constantly changing, based as it is on the co cept of *progress* as both inevitable and inherently good. This typ of production, while ostensibly generating a linear history, actua results in something cyclical, as in nature. By its own definition denies continuity with the past and therefore any possibility building towards a future—it exists only in an impoverishe present destined for rapid replacement. With means havir become ends, progress loses any sense of direction, and objec are produced only to be consumed.

What is constant in things produced as objects merely for co sumption, as Heidegger has said, is their false surface. We a only too familiar with the way in which machines in our time ar enveloped in arbitrarily shaped enclosures: "skins" that, rathe than revealing the forms of the mechanisms underneath, ar determined by criteria related to fashion and market studies. Th yearly "progress" of automobile model changes is the most obv ous example of what is a pervasive reality—a reality that has pr duced a directly synonymous condition in the superficial skins defined literally as "packaging," that envelop contemporary high rise office and apartment buildings. In the same way as with th automobile, the substructures of these buildings are determine solely by technological and economic criteria while the skin respond to the change of fashion, here clearly understood as masking device.

The "readings" of these masking facades (the only manner i which they can be appropriated) are inherently superficial and car only be understood as *information*, whereas human life, and with architecture, is predicated on *experience*. That our contemporary period is called "the information age" indicates the degree of ou destitution. The machines most representative of our time—the television, the telephone, the camera and the computer—are those that deal exclusively with information and therefore deny al but the most emaciated forms of human experience. Indicative o this impoverishment in the realm of experience is the fact that in certain contemporary architecture the real, "present" movements that result from human inhabitation are considered banal, while the implied, "absent" movements involved in a mostly inherited machine imagery are considered expressive. To the degree that the former acts to resist the inflexible demands of optimized tech- nology, the latter must be seen as a concealment or masking of it—as yet another style and, as such, far indeed from any real opposition to historicizing form.

How can mere inhabitation, the simple acts of everyday life, be considered as any kind of resistance? Technological thinking, which believes in the optimum use of universal processes and techniques—the standardization of artificial lighting, climate con- trol, building components, structural systems, construction methods and site preparation—has made it virtually impossible for architects to respond to regional differences in climate,

ography, culture, and construction practices, and to daily
anges in light, weather, and human use. In this sense, the abil-
to move, adjust, vary, and otherwise manipulate architectural
ements, evidenced in most of the following projects, may be
derstood as having a liberative potential arising from its *resis-
nce* to the universal application of optimized technology—but
ly if such operations are determined by the specific nature of the
ace and the human activities that take place therein. Without
perience as the determinant, such operations seem inevitably to
subjugated by utility.

As Adorno has indicated, once under these demands of pure
nctionality, technology can no longer be experienced, only
perated—resulting in our gestures becoming brutal and precise,
ithout deliberation or civility. We are "processed" through the
oving walkway, the revolving door, the escalator, so that even
e simple opening of a door to enter a room, and closing it behind
s, is becoming obsolete: the door now snaps shut behind us
utomatically. The increasing use of air-conditioning and hermeti-
ally sealed windows no longer allows individual adjustments of
ght, air, and view—the swing of the casement window, the broad
ill overlooking the street, and the delicate adjustment of shutters
nd shades are all being eliminated in the production of optimized,
standardized spaces. Irregular sites are flattened by the bulldozer,
liminating all traces of the particular characteristics of the place,
n order that it may efficiently accommodate a maximized volume
of universal space.

Where does this leave architecture? As the Greeks conceived
t, architecture was a unique fusion of form and place. The concept of *boundary*—clearly articulated and concrete—was of essential importance, according to Heidegger; the bounded domain, as
both space and form, allowed the dwelling of man to "take place."
Modern technological thinking, tracing its roots back to Newton,
considers the relation of form and place to be irrelevant: all forms
are the same, object-bodies subject to the laws of science, and
therefore all places are the same, positions measured in relation to
other positions. This kind of thinking thus literally has no place for
architecture as it is understood in its original, essential nature.
Places may thus be understood as "forms of resistance."

Architecture, as the grounding of experience in *place*, has to
do with statics, balance, and the distribution of structural forces in
order to allow this essential rooted condition. Machines, on the
other hand, are essentially concerned with *motion*, as the conver-
sion of energy into work, and may move through space from point
to point: machines may thus be understood as belonging to no-
place. Particular care must therefore be taken when attempting to
conceive of buildings as machines, for machines tend not to be
rooted to a particular location and, as objects displaced into space,
may not be capable of participating in the creation of a larger
order. Herein lies a danger—in using the machine as an inspiration for architecture, the technological thinking that determines the
machine may be used to determine architecture, and the life that
takes place in it.

As determined by technological thinking, the machine is inevitably consumed in *use* and in the incessant *progress* of science
and fashion. On the other hand, architecture as defined by Le
Corbusier involves the *useless* and the *permanent*. The useless

fig. xi. Factory worker.

fig. xii. Pre-fabricated housing.

fig. xiii. Airplane over factory building.

fig. xiv. *Coop Himmelblau: Apartment house, 1983.*

fig. xv. *Plane, Boat, Bomb.*

fig. xvi. *Machinery exhibition.*

and the permanent act to displace utility and progress from the postition as the absolute determinants of contemporary life. Places allow experience, and as Adorno has noted, the power of experience destroys the illusions of progress and gathers the past and future into the present. Collectively, these architectural characteristics also stand against the ultimate "end" of technological progress—nuclear destruction of the human race. In the building of something permanent, the possibility of such annihilation is rejected, and more original human values are given form represented in the world.

It is surely detrimental, then, to the efforts of those attempting to find a *machine-architecture*, that in certain contemporary works the image of the machine is treated almost as an historical "ruin," something pieced together in the aftermath of an implied "inevitable" nuclear holocaust. Architecture has herein become an archaeological process, but one where the elements are artificially aged and distanced—the instant history of a decidedly negative future. Relieved of the responsibility of proposing a constructive vision, these works escape into a hopeless nihilism, not really any different from the pitiful nostalgia for some idyllic past into which historicist architecture escapes. Perhaps the movies are the place for such speculations, but they are wholly inappropriate to architecture and are against the inherent optimism necessary in the act of building.

In light of the above, it is important to remember that machine forms, in and of themselves, are not capable of expressing, much less affecting, any real resistance to determination by technological thinking: the machine carries no meaning. Yet it is this very "neutrality" that makes it most dangerous for us; if we accept the machine as neutral, as something we can use "without thinking," then we risk being delivered over into a type of being completely determined by technological evaluations—*meaning* replaced by mere *means*: utility. As Arendt has pointed out, utility established as meaning generates meaninglessness.

How, as Van Eyck has asked, can architects propose and build *counterforms* for a society that itself has no *form*? The answer lies in attending to the original nature of man himself, which remains always essentially the same. Opposition or resistance to the dominance of technological thinking in contemporary society thus may be attempted throught the revelation of an original relationship between man and technology. Such a discovery is not to be made at only one particular moment in history: the origin of a thing constantly reasserts itself as "the same." Such a beginning might be found in the monastery and the clock.

In the monastery, human life was given a collective order and with it a rhythm of action. As Mumford has noted, the clock, initially acting only to measure the hours, soon became a means of ordering and synchronizing the actions of life, dissociating time from natural and human events. Yet both the monastery and the clock can be understood as having in their original natures the potential for resistance to technological determination. The monastery is first and foremost a bounded realm, a place in its most rooted, permanent sense. As exposed in its courtyard, the earth as *ground* is made the focus and center of man's experience. By organizing the routines of work, contemplation and worship on the basis of the movement of the sun, daily life and the aily cycle of the earth are bonded together in experience. In the *e of the monastery, the concept of measurement remains of *econdary importance: measured space is subsumed in the idea f infinity and measured time is subsumed in the idea of eternity. n the end, the clock must always be adjusted to match that ultimate definition of the hours and minutes—the daily revolution of ne sun through the sky. A clock that is directly motivated by the un, while at the same time acting to define a particular place, vould also bond daily life and the daily cycle of the earth together n experience. Stonehenge was perhaps such a clock.

Technology and the machines that came of it were originally *experimental* in nature—having to do with human experience and *he discovery of the world. But technology has since lost its nature *of being an experiment: a calculating, optimizing, economizing ntention has taken over, and this narrow, utilitarian reasoning now determines the direction of what are still called *research* and *invention*. As Marx predicted, invention has become a branch of business, determined by the demands of production and consumption. Today's technology is essentially devoid of the risk that attends any truly inventive or creative act. This contemporary technology and the machines it produces are unrelated to the human values that motivated the early inventors.

The inventions of these original technologists are, by today's utilitarian standards, arcane, outmoded, excessively gestural, inefficient, and very frequently "failures"—failures as machines but wonderful successes as human endeavors. If these inventive experimental machines are used as sources of inspiration in the creation of contemporary architecture, so as to reground it in experience, it may thereby be possible to *use* the machine while at the same time rejecting the economic determinism and technological optimization that has reduced architecture to such a woeful state of dependency. The machine, one can therefore conclude, can be assessed using values other than those associated with technological thinking; it is by their very *uselessness*, in the progressive technological definition, that these original, archaic machines remain open to contemporary experiment and experience, and why these technological dinosaurs, these extinct species of invention, remain so alive to artistic development. The modern age opened with the intention of liberating mankind through the capacity for invention, for experimentation, for wonder; in our contemporary age of emaciated experiences, the true modern spirit may dwell in an anachronism: the experimental machine.

fig. xvii. Ferry bridge.

fig. xviii. Gliding machine.

It is contemporary man's peculiar tendency to place economic and technological considerations ahead of more fundamental human values. As a result, technology itself has become alienated from us, and the world defined by technological evaluations has acted to alienate us from our fellow men. Our human relationships have progressively degenerated to the mere exchange of information, and our objects are increasingly determined only by their usefulness. Yet at the same time that our relations to people and things are becoming ever more abstract, Adorno points out that the power of *abstraction*, essential to any act of creation, is vanishing.

This situation cannot be acceptable to those whose art requires and calls for the engagement of *making*, a poetic act of

fig. xix. Antonio Sant'Elia: Città Nuova, 1914.

revelation directly related to experience and directly derived fr[...] *techné* as bringing forth into presence. Architecture cannot fo[...] places without the tectonics of making: technology must be [...] engaged. Architecture that is resistant to determination by techn[...] logical thinking may be accomplished through research into mo[...] original relationships between man and technology; throu[...] rediscovery of experimental and inventive technological creatio[...] and through re-engagement of technology and experience in t[...] making of tectonic form. If architecture is to escape from the en[...] less cycles of use and progress, it must place itself in oppositio[...] in this effort the machine offers both danger and hope, for if suc[...] opposition is to be possible, it must come from *within*, from a mo[...] original conception of the nature of technology itself.

Technology is therefore no mere means. Technology is a way [...] revealing...Techné belongs to bringing forth, to poiesis ; *it is some thing poetic...Essential reflection upon technology and decisiv confrontation with it must happen in a realm that is, on the on hand, akin to the essence of technology and, on the other, funda mentally different from it.*
The Greeks conceive of techné, *producing, in terms of lettin appear.* Techné *thus conceived has been concealed in the tector ics of architecture since ancient times.*

—Martin Heidegge[...]

BIBLIOGRAPHY

Adorno, Theodor. *Minima Moralia*, London: Verso, 1978.
Arendt, Hannah. *The Human Condition*, Chicago: University of Chicago Press, 1958.
Banham, Reyner. *Theory and Design in the First Machine Age*, New York: Praeger 1960.
Benjamin, Walter. "The Work of Art in the Age of Mechanical Reproduction," in *Illumina tions*, H. Arendt ed., New York: Schocken, 1969.
Collins, Peter. *Changing Ideals In Modern Architecture*, Montreal: Queen's University Press, 1965.
Frampton, Kenneth. *Modern Architecture: A Critical History*, New York: Oxford Univer- sity Press, 1980.
_____. "Towards a Critical Regionalism: Six Points for an Architecture of Resistance," in *The Anti-Aesthetic*, H. Foster ed., Port Townsend, WA: Bay Press, 1983.
Giedion, Siegfried. *Mechanization Takes Command*, New York: Oxford University Press, 1948.
Heidegger, Martin. *Being and Time*, New York: Harper and Row, 1962.
_____. *Poetry, Language, Thought*, New York: Harper and Row, 1971.
_____. *The Question Concerning Technology*, New York: Harper and Row, 1977.
Husserl, Edmund. *The Crisis of European Sciences*, Evanston: Northwestern University Press, 1970.
Le Corbusier. *Towards a New Architecture*, New York: Holt, Rinehart, and Winston, 1960.
Marx, Leo. *The Machine in the Garden*, New York: Oxford Univ. Press, 1964.
Mumford, Lewis. *The Lewis Mumford Reader*, D. Miller ed., New York: Random House, 1986.
Ortega y Gasset, José. *The Revolt of the Masses*, New York: Norton, 1932.
Pérez Gómez, Alberto. *Architecture and the Crisis of Modern Science*, Cambridge: MIT Press, 1983.
Wright, Frank Lloyd. *An American Architecture*, E. Kaufmann ed., New York: Horizon Press, 1955.
_____. "The Art and Craft of the Machine," in *Roots of Contemporary American Architecture*, L. Mumford ed., New York: Dover, 1972.

x. "A tug of war with a dirigible."

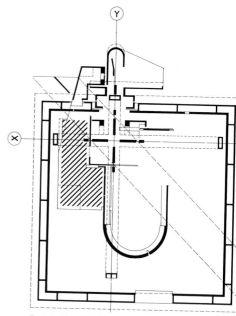

fig. 1.1. Heuristic Structure.

THE (CITY) CONTEXT OF THE MACHINE

In the city, the ironical and most compressive container of our constantly de-centering culture, the strata of the masses are not as distinct as those of the Earth below us. Each subculture, which is part of the continuing readjustment of all people, promotes its own separation, the result being a confrontation with the adjacent subculture. These sociological movements may be seen not only against the context of physical (re)placement but also against that of thinking. Culture, then, must be reflective of an attitude about things or events known as facts. Since scientific concepts are known as facts, the question of the formalization of science, as in the discipline of architecture, is at hand.

The adversarial nature of culture and space (or technicism or technetronics) must endure in and be responsive to our present condition if it is to approach a language of LEGITIMATION. Against the context of the social, technicism bears the imprint of our known cultural precedents while incipiently producing new ones. From this, it is possible and quite practical to believe that the "mechanicality of things" will not continue to progress. And so, the cult of Technicism is elevated to an institution as inevitable as the cult of Humanism.

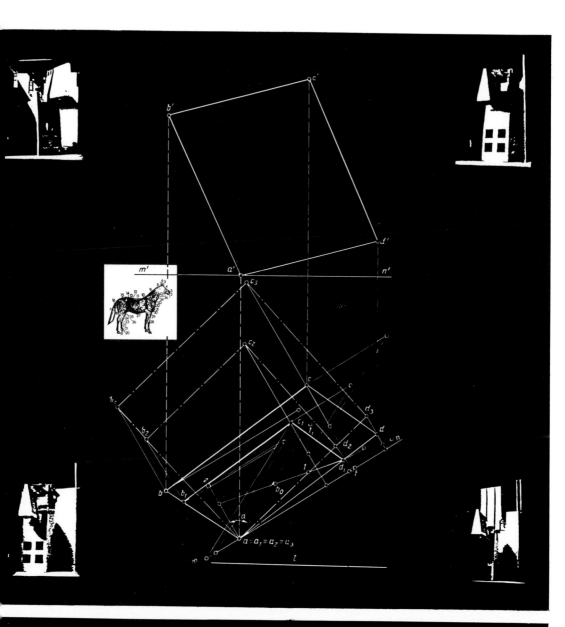

Monastery
New York City No. 8305

The monastery, as a valuable contemporary program, is a refuge from the progressive exhaustion of the metropolis, a place where spirit and thinking are separated from but reflective of the city itself. As a collapsible set of realities, described in a process of scalar reductionism—from the infinite, cartesian world of the city to the smallest cell—the intention is to relinquish our usual mental habitat, to suspend for any length of time the normal processes in favor of a more contemplative one. The monastery is machine as it mediates the energy between man and deity, thus claiming its mechanistic quality by converting energy not into work but into Being.

The site, in the Chelsea section of New York City, is 210 square feet, with 15-foot setbacks from the curb lines. The vertical density of the city rises to a height of 60 feet, a point which the parallel concrete walls (10 ft. on center in plan) align with in elevation. These walls are an abstract form of poche, like a three-dimensional rendition of building solid. The northern- and southernmost walls are virtually blank, with few perforations, while the endgrain of all the walls creates a diaphanous, somewhat filmlike reading of the monastery object situated at the western end of the square.

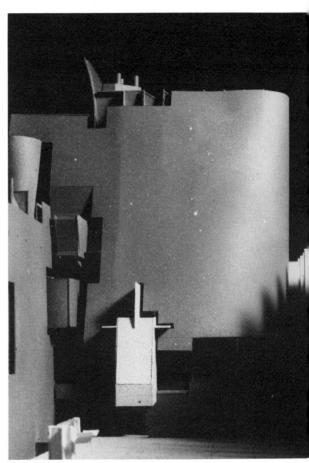

fig. 1.2. Cortile.

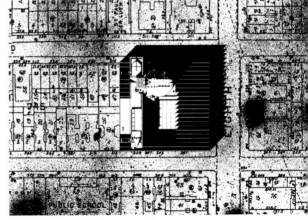

fig. 1.3. Site Plan 8305 Chelsea NYC

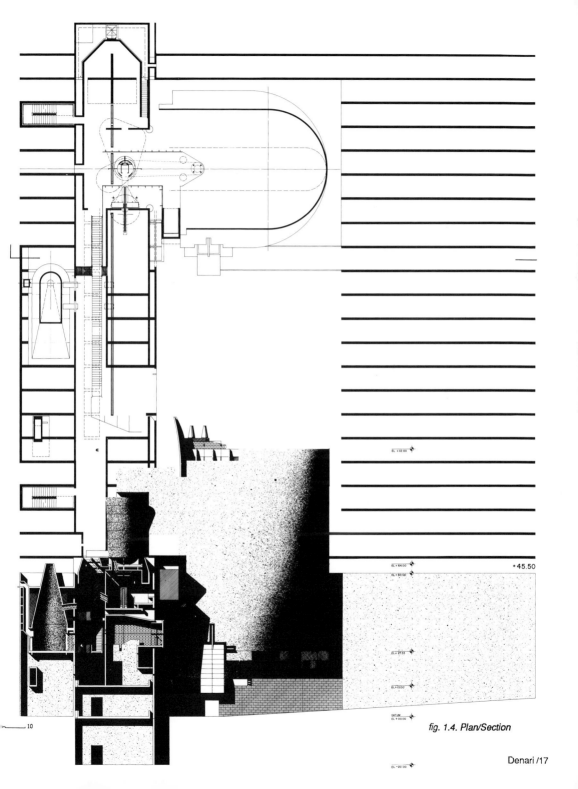

+45.50

EL + 64.00
EL + 60.00

EL + 02.00

EL + 27.33

EL + 15.00

DATUM
EL + 00.00

fig. 1.4. Plan/Section

EL +20.00

10

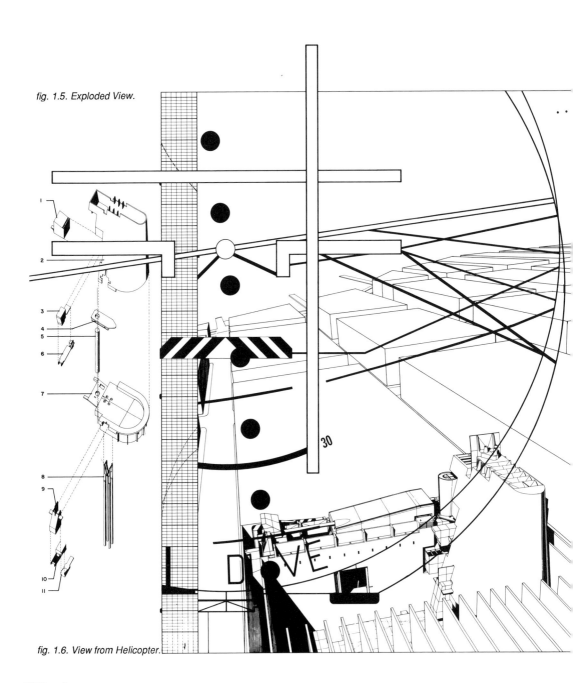

fig. 1.5. Exploded View.

fig. 1.6. View from Helicopter.

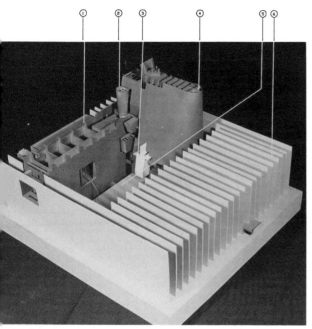

. 1.7. Components.

LEGEND

1. *Assorted Chapels*
2. *Liturgical Vestments*
3. *Entrance/Narthex*
4. *Distorted Cone*
5. *Cortile*
6. *Building [City] Poché*

The machine is an instrument for describing things. At the same time, it expresses its own state. Inherent in the mechanism—that is, the operability of the machine—is an aspect of function that is conjunctive with its self-descriptive function. This is THE ARGU-MENTATIVE FUNCTION, which, according to Karl Popper, is distinguishable from, say, programmatic function. Within the building, there exists a proposition about itself which is communicated externally in the form (or function) of an argument. This argument serves as an expression of the internal mechanism of the building insofar as we accept the challenge of response to this argument—to agree, disagree, or propose an alternative. It is only in this acceptance and human interaction that the argument may find its most useful and powerful state.

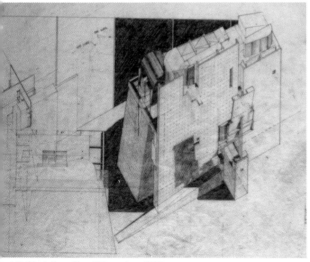

. 1.8. Study for a Rural Monastery 1983.

For the autonomous building to engage the scale of the entire city, to generate vibrations far beyond its own site perimeters, it must somehow embody the essence of the machine: the device in pursuit of liberation. The inherent productive capabilities of the machine must, therefore, always imply a context larger than the thing itself.

Progressive modern mechanisms, from churning pieces of greasy steel to heat-molded plasticine assemblies, are not mere updates on the chisel and donkey. The idea is the same: the desire to reflect ingenuity for efficiency. Thus, the proper understanding or reading of a mechanism is not in its form, but in its intention. Peter Cook has written in the book *Experimental Architects* that "many experiments have incorporated the extra vocabulary of science and technical invention into the rag bag of architectural parts and, in many cases, for good functional reason. But the attractiveness of the objects is appealing, to those who are just looking for stylistic sustenance—and they begin to be borrowed in an indiscriminate way." As such, we see why a general public believes that science has strangled architecture by forcing it into unpleasant, modern forms. The machine is not a RE-actionary tool, it is a HEURISTIC tool. The resonation of its energy and incipient output (in the case of architecture, a production of new environmental meanings) forces an action of positive communication.

fig. 1.9.

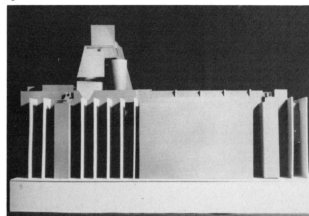

fig. 1.10.

fig. 1.11.

fig. 1.12.

fig. 1.13. Vestment Machine.

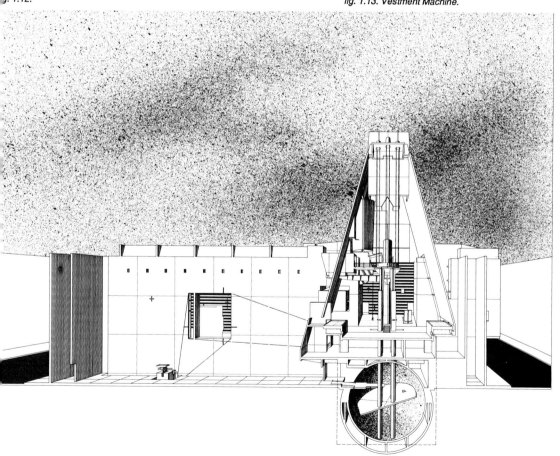

fig. 1.14. Elevation/Section.

Solar Clock
London No. 8602

THE (UNIVERSAL) CONTEXT OF THE MACHINE

The solar day, whose measure of time is based on a single rotation of the earth, is described in a new way via the circumnavigation by this clock machine of the Tower of London. Set upon the outer defense wall of the Tower, riding on a rail-set inserted into the existing structure, the object completes one loop around the site in 24 hours. It is a machine of approximation, a grounding of the SUN in the city of London.

The building envelope is a photon-absorbing, solar curtain wall developed at Stanford University, capable of utilizing 30%

fig. 1.15. Aerial View.

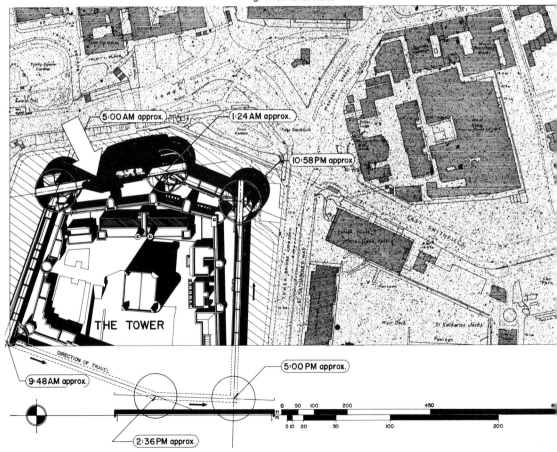

5:00 AM approx.

1:24 AM approx.

10:58 PM approx.

THE TOWER

DIRECTION OF TRAVEL

9:48 AM approx.

5:00 PM approx.

2:36 PM approx.

fig. 1.16.

g. 1.17.

g. 1.18.

g. 1.19.

of the sun's available energy. Thus, the object is indicative of the sun's productive range. Equipped with no numerical ordinances of conventional time-telling, the OBJECT, like the position of the sun in the sky, "tells" us the time only through its relative position in the city context. This building is intended to be so intrusive and reflective that, as a confrontational device, it cannot be avoided.

The first law of thermodynamics describes the origin of all mechanical processes: in the indestructibility of energy, a machine's primary function is, through the generation of heat and motion, TO DO SOMETHING. In the earliest applications of the phenomenon of movement, mechanisms were built in order to measure things. Thus an axiom was developed: time is movement via the consequences of events. Reason enough for the clock to become the device to approximate where we are in the course of these events.

To make a Husserlian description of a collaboration between TIME and the MACHINE is obvious. The machine may be considered as *in* the world yet displaced in turn by its almost daily modification. This continuous, almost self-perpetuating regeneration speaks of the *now* of the machine, describing technology's development as inevitable as time itself. However, the *intentionality* of the machine, as an a priori program, is often misdirected by two major forces: (1) a subversion through proliferation, and (2) invalid intentions.

It is clear that the present condition of the world will not operate under such chaotic terms. New dynamic states of architecture may originate, states that reflect the interaction of a given system with its surroundings. Within this highly potent mixture of discovery and interchange, we are, unwittingly, being subjected to a world liable to control and manipulation. Martin Heidegger has found that the very heart of all technological endeavors relates to the DOMINATION OF NATURE. The violence lurking in all possible forms of communicable knowledge is threatening our understanding of rational enterprises. Therefore, the will to dominate suggests a world of disenchantment in the face of progressive modern mechanisms. ARCHITECTURE demands, the UNCONCEALMENT of this dilemma.

To make an "Architecture of the Machine" is not unlike fabricating a philosophy of science. Peter Cook has written that it might be an "architectural experiment that contains an aggregation of parts that are all the latest of their kind, functioning and mechanistic as an image." But this reliance upon state-of-the-art equipment for discovery is not the point. Architecture is not empirical, the results of building being a representation of imprecisions and uncertainties that persist through given physical laws.

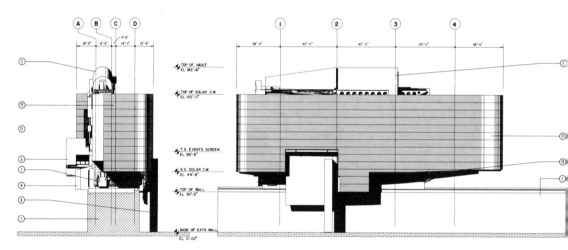

fig. SC3-01 END VIEW

fig. SC3-02 SIDE VIEW

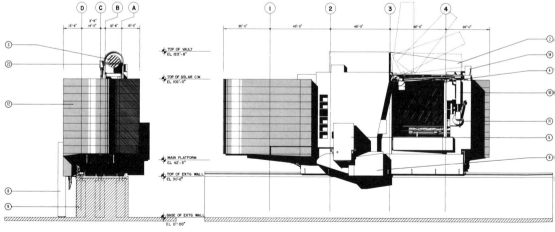

fig. SC3-03 END VIEW

fig. SC3-04 SIDE VIEW

fig. 1.20. Elevations.

The overturning of these negatives produces the clearest possible understanding of the present condition.

Some parts of the world may in fact operate like machines. These (parts) are usually described as CLOSED systems. Reciprocally, the universal or OPEN systems are of the most importance to us. They exchange energy and matter and, more importantly, INFORMATION with their environments. If the world is a big machine, then thermodynamic laws challenge its efficiency by citing energy losses and leakage. These irreversible processes brought about the birth of ENTROPY, still today one of our most confounding phenomena. There is hope in this radical uncertainty: it is precisely this leakage which allows architecture to make order in the face of fear.

Thus a living organism continually increases its entropy—or, as you may say, produces positive entropy—and thus tends to approach the dangerous state of maximum entropy, which is death. It can only keep aloof from it, i.e., alive, by continually drawing from its environment negative entropy—which is something very positive as we shall immediately see.
—Erwin Schrodinger, *What is Life?*

NOMENCLATURE

ITEM	DESCRIPTION
1	OUTER DEFENSE WALL OF THE TOWER
2	OPERABLE COMMUNICATIONS MAST
3	ROTATING LIGHT FILTER
4	HYDRAULIC CARRIAGE
5	SURVEILLANCE BOX COUNTERWEIGHT NO. 1
6	CONTROL PORT COUNTERWEIGHT NO. 2
7	ACTUATOR
8	PUBLIC ELEVATOR ACCESS COUNTERWEIGHT NO. 3
9	INSET STRUCTURE
10	LASER SEQUENCER
11	LASER
12	POINT CONTACT PHOTOVOLTAIC CELL CURTAINWALL
13	CURTAINWALL ADJUSTMENT SYSTEM
14	END FLOAT
15	DIGITAL EVENTS SCREEN COUNTERWEIGHT NO. 4
16	HONEYCOMB MESH MAIN PLATFORM
17	VERTICAL STRUTS
18	MAIN CANTILEVERS
19	LUBRICATION FEED
20	VERTICAL MAINFRAME
21	VIEWING PLATFORM
22	
23	INTERMEDIATE STIFFENER

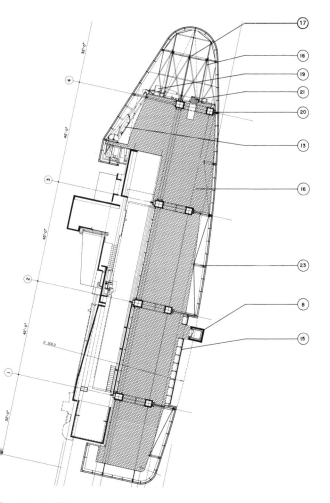

Fig. 1.21. Plan View.

Adam's House (in Paradise)
New York City No. 8407

This project was the result of an invitational exhibition, instigated by the impending removal of the public garden created by Adam Purple, a local folk hero, in favor of city-owned housing. It is located between Stanton and Rivington Streets fronting Eldridge Street in New York's Lower East Side. The program is to provide a maximum number of low-cost housing units while preserving Adam's Garden.

The proposition made is a SLAB-MACHINE that is a hybrid structure of two housing types. The first four floors are typical New York walk-up flats with stairs entered from the street. Above, riding on pilotis, is an external corridor *unité*-type slab. Affixed to the slab as mechanical appendages are cooperative functions such as the game room, laundry, and library. The roof garden is reserved for the expansion of Adam's Garden; the complete object is seen as a wall of observation witnessing nature's productivity.

fig. 1.22.

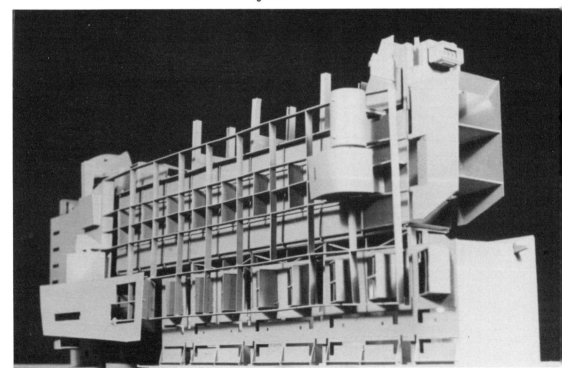

fig. 1.23.

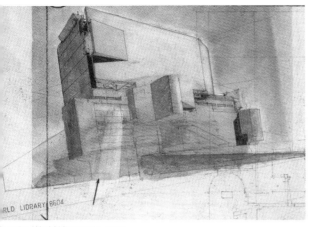

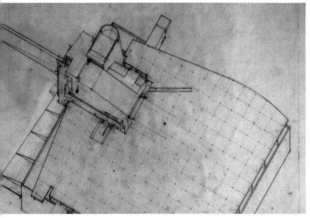

Fig. 1.24. World Library No. 8604.

In these works in progress, the epistemology of the LOGICAL SCOPE OF THINGS is at question: How much knowledge is there and how is it gotten at? How is it confronted and how is it applied justly to a world where we must eat negative entropy in order to generate *new* information structures (like the DNA molecule)? Is scientific inference the (only) basis for generating information? Do buildings then amount to huge pieces of scientific apparatus?

Fig. 1.25. Public School No. 8606.

Fig. 1.26. Communications Center No. 8601.

I know of only one bird—the parrot—tha
talks; and it can't fly very high.

—Wilbur Wrigh
(in declining to make a speech in 1908

fig. 2.3.

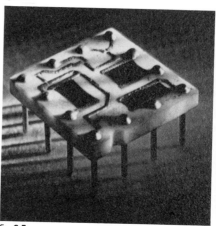

fig. 2.5.

2.1.

2.4.

2.6.

fig. 2.2.

K A P L A N

K R U E G E R

S C H O L Z

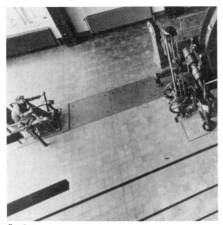

fig. 2.7.

Workstation

A workstation consisting of a 3' x 30' aluminum-Kevlar honeycomb work surface supported six feet above the floor by six bicycle wheels.

The wheels are supported by a steel frame that also supports the seat, lighting, information processing and drafting equipment. The work surface slides laterally past the seat allowing various pieces of work to be laid out simultaneously.

fig. 2.8.

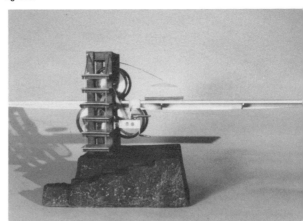

fig. 2.9.

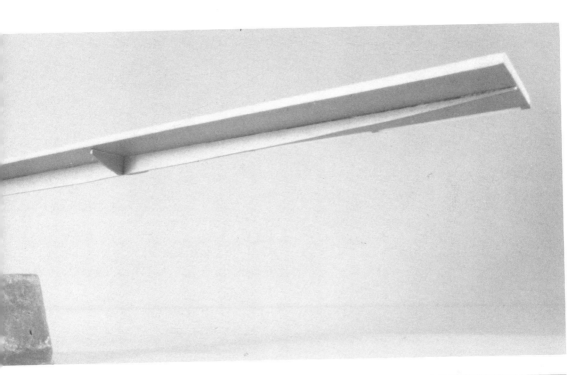

fig. 2.10.

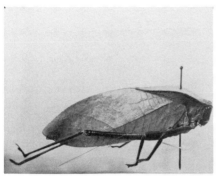

fig. 2.11.

Lamp-Table

This piece of furniture fuctions as both a
table and a lamp. When horizontal, it is a
table capable of seating six people. Its
leaves can be folded up and raised vertically
to become a mobile standing lamp. It measures three feet wide as a table and seven
feet tall as a lamp. Materials include translu-
cent Lexan, aluminum, brass and rubber.

fig. 2.12.

fig. 2.13.

fig. 2.14.

fig. 2.15.

fig. 2.16.

34/ Kaplan, Krueger, Scholz

ig. 2.17.

ig. 2.18.

fig. 2.19.

36/ Kaplan, Krueger, Scholz

. 2.20.

2.22.

fig. 2.21.

Crib-batic

This construction consists of two elements: a child's stroller and its garage; one static, the other dynamic. The static element serves as a flexible spatial enclosure that can be oriented either vertically or horizontally. The dynamic element is a vehicle for stimulation and a tool cart with the capacity to move through space and to present the child with a selection of tools to aid in exploration. These tools, rich in reference to simple mechanics and elemental material, were selected to be especially stimulating to the absorbent memory and intelligence of a three year old.

fig. 2.23.

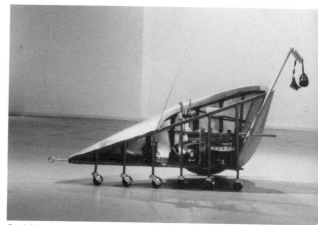

fig. 2.24.

fig. 2.25.

fig. 2.26.

fig. 2.28.

Typing Ball

Wall sculpture constructed of brass, steel, thread, microscope slides, photo transparencies and typing ball. The piece measures 9" x 7 1/2" x 11 3/4".

g. 2.27.

fig. 2.29.

Indoor Machine/220 v.

fig. 2.30.

fig. 2.31.

Chair

The sitter's weight transforms the fabric plane into a sling, pivoting the front frame up against springs in the base.

The chair is constructed of cast and sheet aluminum, chrome-moly tubing, fiberglass cloth, stainless steel cable and hardware.

fig. 2.32.

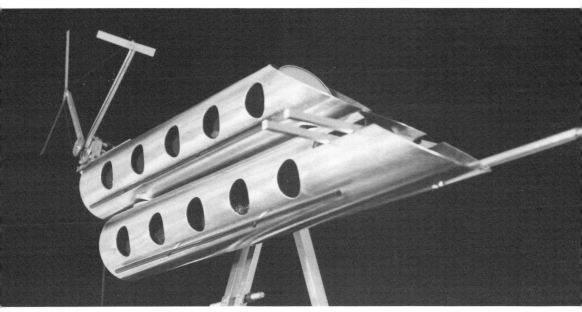

2.33.

Outdoor Machine (Birdhouse)

Aluminum, steel wool, plexiglass and motor(bird).

***What is chiefly needed is skill
rather than machinery.***
— Wilbur Wright, 1900.

Fig. 3.1. "All the splendors of Architecture ever co ceived have been modeled on the little rustic hut. Marc-Antione Laugier, 1753.

ORIGIN VERSUS EXISTENCE AND LANGUAC

Origin is doubly prevented from fulfilling wh it promises: in logic it is defeated by e istence; in language it is defeated by bein named. In both cases its defeat rehearse the impasse of deferral that lies at the hea of architecture: the deferral of uncertainty, meaning, and of value, which creates at th site of the disappointment a richness ar possibility out of which architecture is co tinuously re-invented.

By existing, origin achieves a spatial cor plexity and temporal antecedence that belie its status as limit. As Zeno pointed out, b ginning is always complicated: either there a point of origin or there is not. The only po sible understanding of the *existence* of orig occurs in its evolution, through complicatio to the status of concept. Before undergoir this expansion, it is meaningless; after, cor promised. This paradox can only be avoide by the envelopment of the simple, straightfc ward understanding within a more comple cloud of issues, so that, by virtue of this cor plexity, the limit cannot be found. Circulari is avoided by becoming lost in the clou Like a fractal curve, the concept encloses finite understanding, but the path around it endless.

3.2. The Primitive Hut is the axiom to which classical architecture peals for its pedigree as a natural architecture.

3.3. The hut "empirically discovered" by Laugier is only one possible ori- 1, postulated as a patrimony for a particular architecture.

Primitive Huts, 1985.

Not only is origin defeated in its expansion to understanding but it is lost in being understood. To "grasp the meaning" of it is to presuppose an analogical relationship that already goes against the sense of the concept—in the existence (the signifier) that is graspable, and in the meaning of that signifier, which relates it to other, necessarily prior existences as signifieds. As a further affront, when it is named in this understanding, and so is claimed by language as a sign, origin becomes public. Here origin gains the value found in the intention to communicate, in the desire to *share* it, which again vitiates its originality by implying its value as a commodity. If reading is the act wherein language and existence are united, the active analogizing of things-out-there to the things-in-here (the *creation* of the phenomenal realm), then a concept such as origin, which places itself prior to reading, to perception, to experience, will find itself prior to existence.

AXIOM CONFERS LEGITIMACY CONFERS AXIOM

As with all idealist dilemmas, it is when the concept is elevated into the public realm that it becomes embarrassing. The embarrassment of origin, of limit, is the idea of axiom. The axiom is born from the repression of the absurdities that attend the defeat of origin. The axiom provides an answer to the dilemma of beginning: if *ex nihilo, nihil fit*, then there must already be something with which to begin. The axiomatic beginning is found in "self-evidence": the chain of reductions that arrives at the axiom compels it to be the limit, and thus the beginning, because it is so basic as to be "beyond proof." The preexisting standard of proof, which is the preexisting system the axiom is invented to found, thus negatively defines its relationship to its most basic element. The axiom is recovered, however, through the artifice that arrests the endless deferral of origin, turning the negative structure of the definition into the positive, which legislates it to *be*. From this provisional origin proceeds a post-rationalized evolution which miraculously explains the context that legislated the axiom. Given that context, the intention behind these efforts disappears into nature, and with it the "post-" which undermines the rationalization by remembering its provisionality. The axiom is the sand in the oyster—necessary to begin the process, but, if the process succeeds, swallowed during its course and denied in the pearl.

The oyster creates the pearl in order to protect itself from the externality represented in the grain of sand. It buries this foreign body under layers of nacre—that material which forms the shell of the oyster and protects it generally from the externality of Otherness. The pearl is considered to be one of the world's most ancient symbols of perfection, increasing in value with the increasing purity of its nacreous cloak.

EXTERNALITY BELIES AUTHORITY

What is cloaked is not so much the sand itself as the implications of its absent presence. The value inherent in the heuristic character of the axiom is the residue of externality it must shed if the system is to believe itself; the intention behind the axiom belies its figuration as an irreducible truth. It is only as such a truth that it may pretend to the authority of the one and only reality. Thus, from within this reality the axiom appears as the absolute; the implication of a (necessarily) contrary meta-position is forgotten in the aura of this limit.

It is more obvious in language than in any other system that the obverse face of the invoked absolute is Arbitrariness. In language, the absolute, or axiomatic, limit is the words that compose the system; they delimit (reduce) reality for the particular language. This inseparability of the absolute and the arbitrary is evident to anyone outside the particular system—to a speaker of a different language, for example—but the efficacy of language depends on both the transparency of the word to the non-word that it signifies and the invisibility of the distance between them, which gives the sense of congruence that makes the connection seem necessary. Thus the shock of de Saussure's revelation that the signifier is completely

Fig. 3.4.

Fig. 3.5.

Fig. 3.6. *The obvious heuristic character of the primitive hut's "discovery" provides a license for other discoveries to be made.*

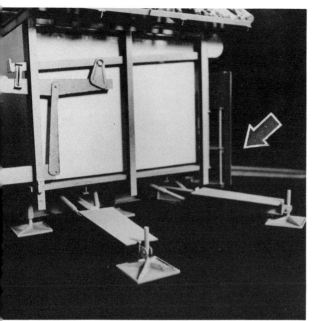

Fig. 3.7. The primitive hut should not be celebrated as a link to nature, but as a step to man; as the first building, not the last tree.

Fig. 3.8. The hut is not the font of nature but of artifice; it is not a natural thing but an electric water-cooler.

unmotivated, that is, arbitrary.

Mathematics, which supplies the particular sense of the "axiom" we are using, was founded upon the faith that the scrim of language could be made to disappear into absolute congruence with reality. Always held up as the standard of exactitude and certainty, mathematics bases this standard on the exactitude and certainty of limiting condition represented by axiom. Based essentially on this sense of the absolute, it is not surprising that mathematics was the first to discover its limitations as a limit. It did this gradually; as mathematics evolved from a concern with "observable reality" to a concern with possible realities beyond appearance the "absolute" was transferred from correspondence to coherence, from a standard of consistency with external reality to a standard of consistency within the workings of the system itself. This move away from Nature was taken as the final step by Gödel, who showed that even internal consistency, a concept derived from nature, could not be proven within the artificial system. This demonstrated that the axiom was a useful heuristic device but could not be validated without recourse to the externality that invalidated it.

THE ORIGINS OF ARCHITECTURE

This tottering concept of the origin is the fulcrum on which architecture teeters. In its instability it is the source of the latter's own expansive in-definition. The coordinates of architecturality, whether art and utility, design and building, ornament and shelter, communication and being, or any of innumerable other oppositions, always locate architecture in the area marked out by their battle for precedence, for the status of origin.

The beginning of architecture is complicated by architecture's simultaneous existence as abstract institution and physical fact. Both claim the origin for themselves; many stories have been told to promote one over the other. The primitive hut story suggests, for example, that building preceded architecture—that there was a hut which became architecture—but the story betrays a prior sense of architecturality by which the hut is recognized. Which came first: the hut or the idea? We recognize that the story is a myth, retro-figured into the history of architecture to provide an origin from which the present

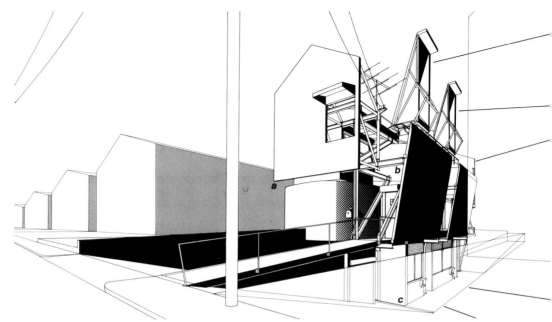

Tract House, Manhattan Beach, California, 1986

Fig. 3.9. The machine is second nature, connecting man to nature as enhanced vernacular—giving form to man's understanding of natural forces

condition could be logically, scientifically derived. The story illustrates an atemporal hieratizing of the opposition-couple to give the effect of an atemporal origin. History aside, there is at the site of architecture an obliquity between building and architecture, presence and absence, *text and supplement*, function and art, the necessary and unnecessary. The deep ineffability of architecture-ness stems from the instability of this ordering, from the constant reversal of these terms and the attendant impossibility of saying architecture with either term alone.

PURPOSE AND ARCHITECTURE AND ORIGIN

In this battle for precedence, one term commands originary status by casting the opposing term as inessential—by becoming the ground condition from which the other differs as an addition or supplement. The opposition between architecture's physical presence as a building and its existence as an abstract

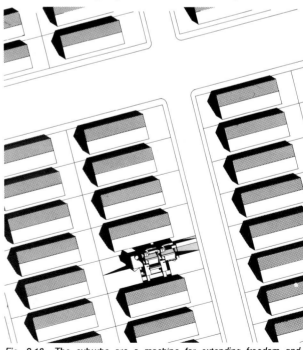

Fig. 3.10. The suburbs are a machine for extending freedom and sovereignty of the individual to an age that has lost the frontier.

g. 3.11. The machine shed the politicizing of technology through its poten-
l for the support of individual, subjective experience.

institution defines the paradigmatic axis of this geometry of supplements; the syntagmatic axis is defined as the two worlds manufacture the sense of necessity that will determine the order of precedence along the pardigmatic. In each case, the character of essentiality being strived for is related to purposefulness. Architecture's difference from building, and its status as a supplement, hinges in each case on the lever of Use. When building is defined by function, architecture is seen as an addition or superlative marked by its lack: Utility is the dominant term. Further, utility is seen as ground condition for determining meaning—including that which art, as a purposeful activity, claims to offer. If a supplement is the sign of the presence of grafted texts/arguments, then in this case the text of art is grafted onto the body of the building as architecture.

But the idea of supplement is also understood as that-which-makes-up-for-deficiencies: "...supplementation is possible only because of an originary lack." Architecture would not be continuously re-invented if it weren't felt that building itself was somehow incomplete. This ordinary lack is not *in* or *of*

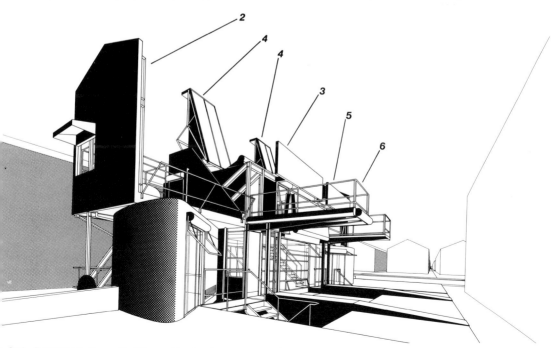

. 3.12. Reference to the world of the machine is reference to the only sig-
cant contemporary reality.

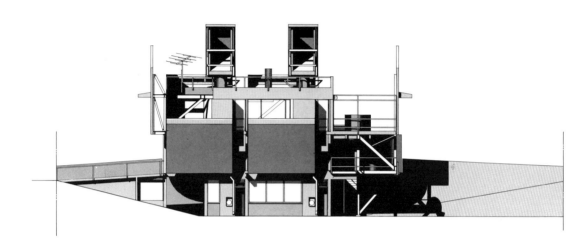

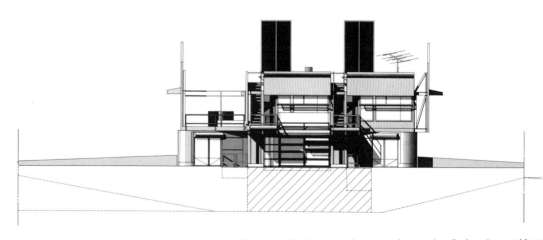

Fig. 3.13. The frame we place around external reality in order to subject it, to cast it as a projection of ourselves, is mechanical.

3.14. Architecture is expressive excess seen against a context of use.

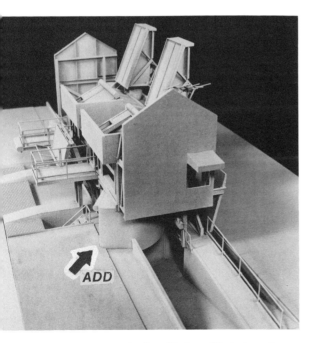

ADD

3.15. The human concerns implicated in the architectual equation are st truthfully and expressively figured by the structure of mechanality.

the building, however. It is rather the building's necessary (objective) distance/-difference from the subject—a distance that can only be overcome by the supplement of analogy, or reading (which, of course, still leaves the actual Other ever derived, deferred). The projected selfness that constitutes reading masks the object as it frames it into a particular reading set. The supplementary concept of architecture becomes the means by which we understand the idea of building itself; it becomes the reading of building. This occurs not only through a dialectical understanding of building as a pre-architectural or non-architectural condition, but also through architecture's structuring presence as the frame or context which isolates for us the territory of building. Those central issues like structure of function, "which appear to explain everything"—by which we label the concept of architecture as an inessential superlative or enhanced "supplementary" condition—are themselves ineluctably caught up in the contextual net that determines their own reading. Their centrality is not given or immediate: it is finally not of the building itself, but of the *architectural* experience of the building.

This attempt to establish the precedence of either utility or art is condemned to the ambiguity that plagues origin in general: assigning absolute priority to an entity always admits the paradox of the endless deferral of priority.

ORIGIN(S) IN MEANING IN ARCHITECTURE

The architectural experience of a building is the expectation that the building be meaning-full or expressive of some intention—intention here understood as the subjective face of the difference/distance that traces out the figure of meaning; it is the will to bridge this distance. With what meaning, though, is architecture full? Most basically, perhaps, architecture communicates its acceptance of the projected mantle of personification. Communication, as the space between the terms, is only possible between essential-others, by mutual, projected reflection. The communication *link* between subject and object is an objective fiction, a bridge cantilevered from either shore but never completing itself in the middle. We "know" by projecting an analogy of what we *do* know—ourselves (interior)—onto the other (exterior).

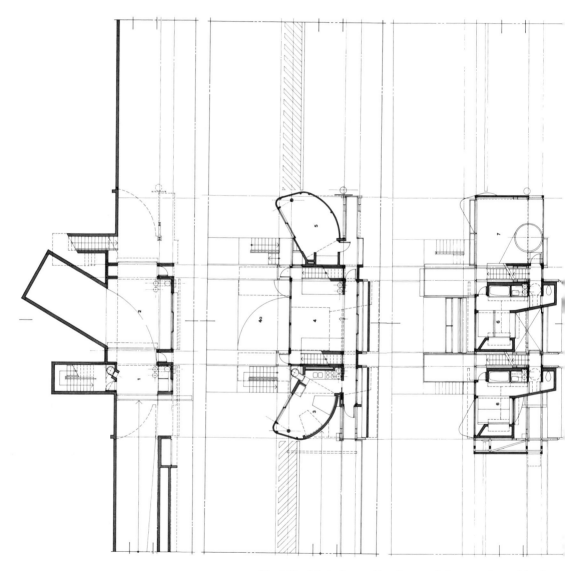

Fig. 3.16. It is in the machine that use finds expression and is elevate through this necessary excess into cognition as significance.

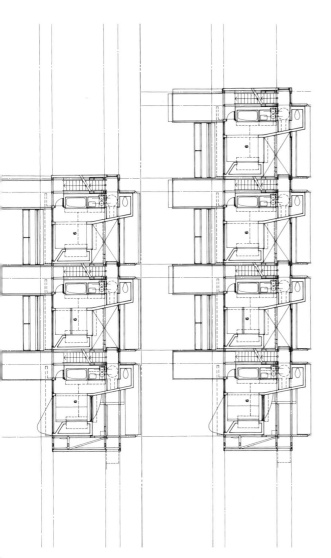

The act of Reading is the conscious flipside of the unconscious act of projection. The effort put forth by the brain in this constitutive reading activity translates into the feeling of the vibrance, or plastic being-there, of the object so read. This equation of cognitive effort with the brilliance of the effects perceived allows translation among the various sorts of effort and explains as well the energy released by acts of repression.

Origin is the key concept in the structure of meaning. Within architecture this key works two mechanisms, formal reference and compositional development, which relate to the two senses of origin as source and as beginning of existence.

REFERENCE

The formal referent is the architectural axiom that sets the context *ex nihilo*. It prescribes meaning through the invocation of a preexisting world of meaning, which directs the reading. The implications of externality inherent to the axiom generally stipulate for architecture that this world of meaning be borrowed from outside the field. The paragon of architecturality is itself empty without an object to frame. By itself it has no presence: it is a verbal noun, an attitude; it has no internal ability to generate form out of the void. Form must therefore be supplied, the void filled by a paradigmatic importation. Along with the form may come its whole syntagmatic chain, and the rules for operating it; or it may be imported clean of its associations, as a misunderstood or reappropriated fragment. In each case it becomes an origin/axiom, a source or signified for the act of reference that constitutes the plastic semantic utterance.

Nature was the referent of classical architecture. The natural world was the source for what the architectural object was supposed to represent, and its mythology was the source for the concepts the object was supposed to express architecturally. Architectural meaning, as opposed to more general cultural or more specific topological meaning, is derived in this case from the space opened up by the *factoring* of the natural world, from the value contained in the paradoxical *artificiality* of this nature, assimilated into the object.

COMPOSITION

Within the specific architectural object, meaning is generated by the compositional development (or "designing," if "composition" has become too specific to certain practices isolated now in history) of the object from an original state. In this case the reference is temporal and syntagmatic—not cultural, cross-disciplinary, or paradigmatic—and it finds its context wholly within the process by which the object manifests its specificity. The origin understood in this way relates to the end product of this process (the completed building or design) as the simple, generalized beginning of its existence, as that grain of sand which, through the agency of the oyster, eventually becomes a particular pearl. The simple house/temple form is the general origin which classical compositional activity transforms into all the particular objects that populate its universe. It is the origin which Laugier invoked as the gateway through which the Forms are translated, via nature, into reality, and which may be seen, vaguely and imperfectly, in each of these objects.

The compositional design process creates a meaning whose principal attribute is elaboration: the finished design contains within it the seed from which it grew, and meaning develops with the design's growth beyond this seed. Reference is thus contained within the object and its own history, rather than being a bridge to a world of external meaning. Thus the syntagmatic process of development is in a certain sense hermetic. Once the process has run to completion, however, the system that has unfolded during the process becomes closed by the completion of the design; this closure immediately suggests the externality that explodes the privilege accorded, *post hoc*, to the seed. The apparent necessity of the particular origin, compelled by the origin's visibility in the finished product, becomes subject to Gödel's

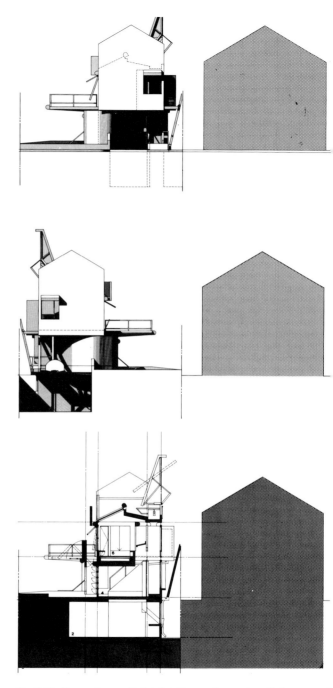

Fig. 3.18. The zealous understanding of pervasive mechanality is tempered by a zealous awareness of the heuristic character of axiomatic origins.

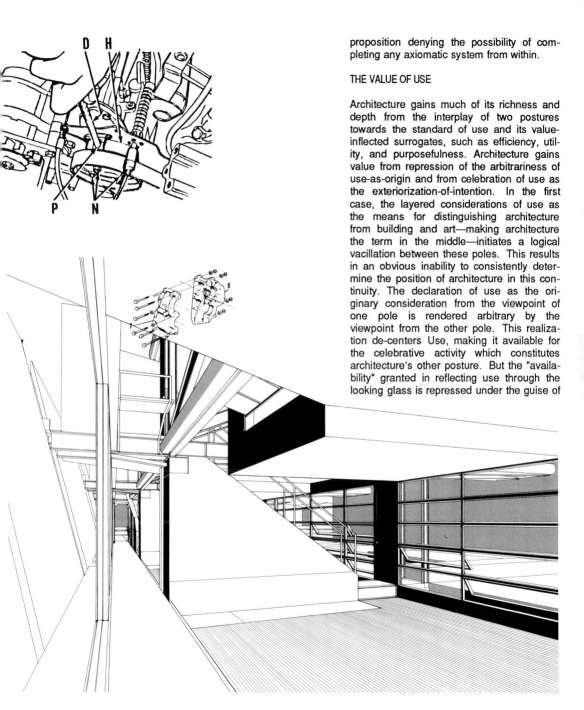

proposition denying the possibility of completing any axiomatic system from within.

THE VALUE OF USE

Architecture gains much of its richness and depth from the interplay of two postures towards the standard of use and its value-inflected surrogates, such as efficiency, utility, and purposefulness. Architecture gains value from repression of the arbitrariness of use-as-origin and from celebration of use as the exteriorization-of-intention. In the first case, the layered considerations of use as the means for distinguishing architecture from building and art—making architecture the term in the middle—initiates a logical vacillation between these poles. This results in an obvious inability to consistently determine the position of architecture in this continuity. The declaration of use as the originary consideration from the viewpoint of one pole is rendered arbitrary by the viewpoint from the other pole. This realization de-centers Use, making it available for the celebrative activity which constitutes architecture's other posture. But the "availability" granted in reflecting use through the looking glass is repressed under the guise of

Fig. 3.19. The process the machine manifests is itself a manifestation of a program of service to humans.

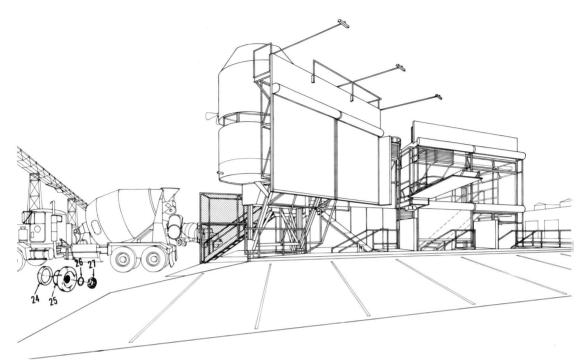

Administrative Facility, Cement Batching Plant, Oakland, California, 1986.

Fig. 3.20. The machine is man's answer to his alienation from nature, not its cause.

the necessary for which it is celebrated: the active repression of arbitrariness enhances the brilliance of its celebration as irreducible necessity.

The value that tradition granted use as the irreducible necessity binding architecture to building and separating it from art is measured here in the same way that meaning is created—in the distance accumulated and destroyed by the object as it moves between these poles. The particular sense of the necessity of use that throws architecture into relief from art as a species or condition of building is the same as that which, from the opposite direction, distinguishes architecture from building as a form of art. With the structure the same, the effects, meaning and value, become interchangeable.

When function is de-centered from necessary originality in deference to meaing, the objective defers to the subjective. Historically, "Function" has become to refer to the

Fig. 3.21. If efficiency is plotted against history, the excess which makes meaning visible lags necessarily behind the available technology.

Fig. 3.22. At the scale of human engagement in the world, reality is inescapably mechanical.

Fig. 3.23. The classical system clothed a far more efficient concrete system. Similarly, the computer age will wear a mechanical face.

objectification of use and by extension, of the measure of use—man. Man objectified is man alienated, and his position in the cosmos shifts radically as its dominant value is engineered in terms of efficiency. Efficiency lies on the same curve as the congruence between object and its purpose, its transparency to itself and its invisibility to man. Efficiency is exactly counter to the excess that could anchor the bridge of meaning and provide the mirror whereby man might be able to find himself in the object.

In favoring efficiency, the architects of the modern movement were led to consider the technology of mass production as the most refined example of this value-curve and to prefer forms which expressed the objectivness of this process. Their efforts, however, illustrated the paradox of *expressing* efficiency: expression, or the opaqueness that creates the distance of meaning, is contrary to the movement of efficiency, which, of course, is towards *closing* this distance. Consequently, the architect is forced to sacrifice efficiency in order to make it visi-

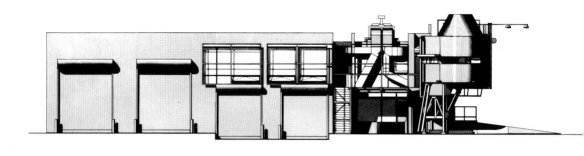

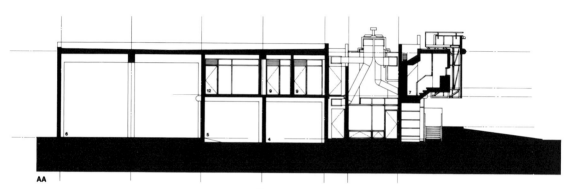

AA

Fig. 3.24. Because of the overtness of its programmatic relation to man, th machine is more engaging and legible than the products of convention.

ble, in order to express it. Those architects who declared efficiency as their standard lost expression, and their work became sterile. The better architects, like Le Corbusier, recognized in the sacrifice of efficiency the doorway to man—and subjectified the understanding of function so that its visibility was heightened (to the point where the critics would claim, obliviously, that Le Corbusier's buildings were not nearly as "functional" as they pretended) and the forms became engaging rather than alienating. In these works, experience supplants function as the standard of judgment. When experience becomes the *telos*, function is restored to the lower-case form of its originary status; it becomes visible again as an intention directed to the service of man, rather than to itself as an end. When the end is man, he is visible in his works as both the true originary condition and its agent. By the same structure that equates the force of repression with the

Fig. 3.25. The machine fades to invisibility with its increasing pervasiveness Because of this, its reality as our reality has not been recognized.

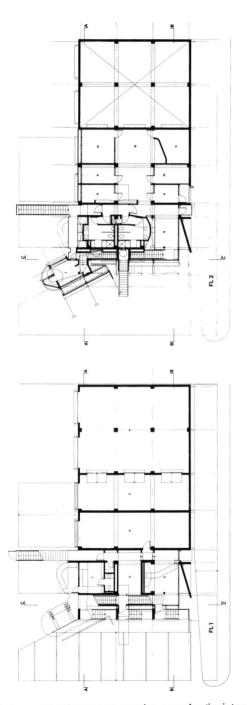

strength of its effects, and increase in the degree of alienating functionality tamed by this subjectiving design process—*in the name of expressing this functionality*—acts to heighten what can only be called the Boss-ness of the object.

. 3.26. The machine that supports experience over function is transparent the human whose intentions it manifests.

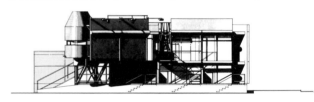

Fig. 3.27. The machine is the same kind of thing as a building; it is not asked to look like something else, but to develop its own expressive potential.

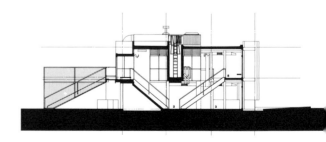

Fig. 3.28.

Fig. 3.29. The machine is not an agent of alienation to be feared; as the
definitive product of his will, it is a signifier for man.

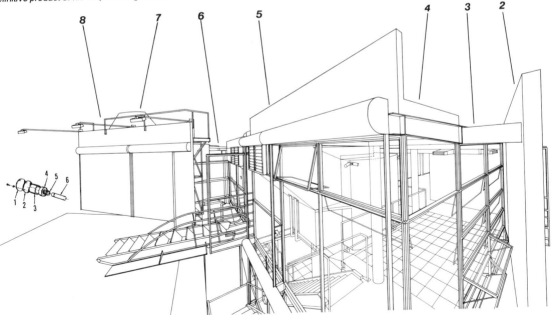

Fig. 3.30. The processual reality of the computer is merely the miniaturiza-
tion beyond comprehension of mechanical choice.

fig. xxi. Iako Chernikhov: Studies for a machine architecture, 1928-1931.

This unexpected return to constructivism is a mixture of many points of view. Of the various movements that come to mind, one thinks on the one hand of British "high-tech," John Hejduk, Lebbeus Woods, and, as incongruous as it may seem, of Charles Simmonds. On the other hand, one may also detect, more conventionally perhaps, the influence of Zaha Hadid, Rem Koolhaas, Bernard Tschumi, and Steven Holl. One may rest assured then that this is a *répétition différente*, as Barthes would have it, and that it is not just a repeat of the heroic constructivism of the 20's. However, certain affinities with that time seem to persist, above all, the dadaist spirit to which Russian constructivism always had a close connection, evidenced here in the absurd machines of Ken Kaplan, Ted Krueger, and Chris Scholz, and in the strange "high-tech" hut created by Wes Jones and Peter Pfau.

All in all, this work can be seen as a kind of "ruined" constructivism, a constructivism that has been reassembled out of the detritus of the modern world. Poetics aside, this is hardly a movement that believes in itself in the way that the first constructivism did. For these young machinists know, even if they do not declare it, that they have little chance of realizing these works. They are, therefore, relatively free from the utopian illusions of the original constructivists. Their works imply that modern culture is now as much destroyed as traditional culture and that neither modernism nor historicism is truly available today as a point of departure.

They are, however, passionately committed to craft, or rather to a constructed mode that, far from being folkish, assumes a tough tectonic form that borders on a kind of "street-art." In this sense their work is particularly alive and exploratory. It might even be classed as experimental to the extent that it represents a tentative but rigorous effort to regenerate a lost syntax. Thus, the mega-high-rise slabs designed by Jones and Pfau are to be valued more for the tectonic vivaciousness of their *brise-soleil* and articulated balconies than for the questionable viability of their implicit spatial and social urban order. Here one is returned to the Vkhutemas and to the Russian Structuralist interest in adducing an Edenic formal language.

The most conceptual (one might even say the most somber) of these constructivists, namely, Neil Denari, posits a much more disturbing world than anything imagined by his peers. His monastery, for instance, is a non-place *par excellence*, comparable to John Hejduk's *The Cemetery of the Ashes of Thought*. This is surely the darker side of the apparently innocent exuberance of this movement, a side that seems prepared to come to terms with a modern world which, in the last resort, has to be seen as a domain of anxiety and pain. However, this more reserved side does nothing to diminish that which is most positive in all this work, namely, the effort to ground architecture once again in structure, craft, and the poetics of construction, rather than in the gratuitous aestheticism of abstract form.

fig. xxii. Kaplan, Krueger, Scholz: Work Station.

fig. xxiii. Wes Jones, Peter Pfau: California Unité.

fig. xxiv. Neil Denari: Monastery.

fig. xxv. Shaker stove.

fig. xxvi. Airplane wing assembly.

MICHAEL SORKIN

Le Corbusier's most titillating turn of phrase was his labelling of house as a *"machine à habiter."* Even now, the conceit seems provocative, going too far, collapsing domesticity's joy and refuge with the alienating grind of regimenting technology. Never mind that to Corbusier's major commitment to his formula was the social engineering of the Ville Radieuse; it is a fault line that pervades architectural practice, the question of the degree to which the inherently mechanical character of building is admissible expressively, the abiding conundrum of functionalism.

The first film I can remember seeing is Disney's *Twenty Thousand Leagues Under the Sea*. What remains most indelibly in my mind is Nemos's sub, the Nautilus, all craggy plates and bolted fittings, the apotheosis of late-nineteenth-century tech. But it was tech with a twist. Captain Nemo occupied a cabin that was a compendium of Edwardian excess, plush red like a brothel. Nemo was a premature postmodernist, straining to domesticate tech before could do the same to him.

I read last year that Steven Jobs was building a big new house out in California. Certainly Jobs, with his Apple Computer millions can afford any sort of house he pleases. His choice? A scrupulously reproduced Victorian. There is a twist here, too: the mansion is wired to the hilt. Behind the perfectly carved, ersatz antique wainscoting glow miles of fiber optic cables leading to an array of outlets for the latest electronics. It is an inverted Nautilus, tech relegated to convenience, banished from form-giving, a mobius of dissimulation.

Naturally, there is no suasion readily available for locating architecture's prosody in its material condition, the *minima moralia* of its enabling science. We live in a time when the incitement to remember overwhelms the imperative to invent. The lush and fulfilling promise of the technical is indexed by a profession that prefers to wallow in the predigested certainty of recombination, the endless manufacture of fresh freaks from a familiar gene pool.

Happily, though, there are those who when the Concorde streaks overhead turn their eyes upwards, who find the idea of a formal language that is rigorously testable, that chafes at constraints, to be the zone of exhilaration and prospect. A jumbo jet, full flaps, gear extended, is more beautiful to them than this week's beach house. Not because one is a machine and the other is not but because the 747 is newer and truer, its research more scintillating, and its ambitions sky high. Architecture's inevitable mechanism must, like the repressed, eventually return. The fulcrum for its liberation is the poetry of its purpose.

The transit of a nail through a plank, the compression of a column of steel, the tense stretch of aluminum skin, the modulation of a breeze, the framing of a view, the pleasurization of a place form the metrics of the built machine. This poetics is not applied but indigenous, not borrowed but discovered, not timeless but temporal, not arbitrary but directed. Architecture's return to the machine—in the full dedication and richness of its possibilities—holds the hope for architecture's future as both servitor and art.

*'ef in the omnipotence of technology is the specific form of
ɪrgeois ideology in late capitalism.* —Ernest Mandel

ɪef in the omnipotence of technology: either total faith or total
lism, either blind trust in technology or passive resignation to
totalitarian logic. Both these positions must be rejected—but in
ɔr of what? Technology is not autonomous. Indeed, only in
time has technology so dominated the social; that is, only in
· capitalism has Marx's vision of a society in which "all the sci-
·es have been pressed into the service of capital" become true.
ɪay, technology is no longer distinct from science; there is
ɪead, "technoscience" (Lyotard), the instrumental integration of
·earch and development, knowledge, and power. This system
des definitions on all sides: the difference between body and
-body, the difference between life and death, the difference
ween nature and not-nature. This is real deconstruction—
cticed not on literary texts or works of art but on our very bodies
I environment. In this type of deconstruction disciplines like
hitecture are eroded, its terms—body and space—are
ɪsformed by new machines and speeds. Given this practical
ɪonstruction of architecture by techno-science, do we really
uire its theoretical deconstruction in the academy?
  Is there any possibility for a resistant architecture? Some con-
ɪporary architects at least address the simulacral nature of our
ɪtemporary urban world—but without much criticality. Just as
plug-in architecture of the 1960's abetted the ideology of con-
ner culture, so too this "techy" architecture of the 1980's abets
ideology of post-industrial society; it neither contests this ideol-
·' nor exposes its contradictions but presents it as an experi-
ɪed totality.
  But if an information aesthetic is no grounds for a resistant
hitecture, one cannot merely resort to the machine aesthetic—
ɪer in its modern guise, in which the machine informed the
hitectural structure, or in its contemporary guise, in which the
chanical is a mere motif. The latter is significant as a symptom,
it suggests not only that we are beyond the modern machine
·, but that the machine is now treated as a romantic ruin.
  How do we get beyond this metaphysics of representation (that
hnology, after all, has eroded)? The first step in developing a
itics of techno-science is to think our way out of any historicist
erminism or imagined necessity regarding the technological.
ɪre is no neat line of modes of production; there is only the con-
ual development of technolog*ies.* And the exposure of this his-
·' opens up a way, both to resist the dominant techno-logic of
First World and, to avoid the romanticism of the non- or less-
ɪnological in other worlds. On a practical level, artists, critics,
I architects might articulate the contradictions between our
·n techno-social paradigms (between, say, the spectacular
Jd of the car and TV and the informational network of the com-
ɪr). Once exposed, these contradictions might renew a whole
ge of positions, each for a specific context or conjuncture.
ɪh a revision might even allow for "convivial" uses of the tech-
ɔgical and—who knows—maybe even utopian uses.

fig. xxvii. Robots on auto assembly line.

fig. xxviii. Iceboat.

SOURCES OF PHOTOGRAPHS

All other photographs courtesy of architects.